Summary and Analysis of

THE IMMORTAL LIFE OF HENRIETTA LACKS

Based on the Book by Rebecca Skloot

This Worth Books book is based on the 2011 paperback edition of *The Immortal Life of Henrietta Lacks* by Rebecca Skloot published by Broadway Books.

ISBN: 978-1-5040-4673-2

Worth Books
180 Maiden Lane Suite 8A
New York, NY 10038
www.worthbooks.com

Worth Books is a division of Open Road Integrated Media, Inc.

Contents

Context

On October 4, 1951, Henrietta Lacks, an African American mother of five from Virginia died of cervical cancer at Johns Hopkins Hospital in Baltimore, Maryland. While beloved and mourned by her family, no one could have predicted that Henrietta's death would change the course of medical science. That's exactly what happened, though, when scientists who had extracted cancerous cells from Henrietta's cervix—without informing her or her family—found that her cells were capable of reproducing indefinitely in the lab. This first line of "immortal" human cells would be used in research ranging from the development of the polio vaccine to cancer detection and treatment, and learning how human cells responded to zero-gravity conditions.

While scientists have long been familiar with the extraordinary HeLa cells as they are known (from **He**nrietta **La**cks), few knew anything about Henrietta, her family, and the effects that her cells had on their lives until Rebecca Skloot's book was published in 2010, nearly sixty years after Henrietta's death.

After more than a decade of research—and making her way past a great deal of initial resistance from Henrietta's widower and children—Skloot reveals for the first time the deeply personal story of the woman whose cells became one of the most important tools in medicine, and the family that was ignored and forgotten by the scientists who harvested her cells without her knowledge or permission.

Though there is no evidence that the doctor who originally harvested the cells or Johns Hopkins ever received direct profit from the sale of HeLa cells, many biotech companies have; and none of the profits were shared with the Lacks family.

The Immortal Life of Henrietta Lacks is a pivotal exploration of the complicated relationship between race, science, and medical ethics.

Overview

In 1951, after more than a year of experiencing pelvic pain and abnormal bleeding, Henrietta Lacks, a young African American mother of five, is diagnosed with cervical cancer at Johns Hopkins Hospital in Baltimore. Hopkins doctors treat her with surgery and radiation, but her cancer continues to spread and she dies within a year of the diagnosis, leaving behind her husband and children.

During her treatment, without her knowledge or consent, doctors remove a sample of Henrietta's cancerous cells. Hopkins's head of tissue-culture research, Dr. George Gey, has been trying to grow a line of immortal human cells that reproduce indefinitely to be used for research that cannot be done on live patients. To Gey's surprise, Henrietta's cancerous cells do not die within a few days, as all his previous samples have, but instead continue to multiply at a rapid pace.

While her family struggles after her passing, Henrietta's cells are on their way to notoriety in the scientific community. George Gey sends samples of the incredible immortal cells to labs around the world where scientists use them in all kinds of research.

It isn't until 1973, two decades after Henrietta's death, that her family learns her cells are still alive. Her husband and children feel confused and angry that they were never informed about Henrietta's cells or compensated for any of her contributions to science.

When science journalist Rebecca Skloot first tries to talk to Henrietta's children, she faces a great deal of resistance. They feel they have been taken advantage of by the media and scientific community. Once Henrietta's daughter Deborah finally agrees to talk to Skloot, they embark on an incredible journey together to learn about Henrietta's past, her incredible cells, and the lengthy struggle for patients to gain the right to informed consent for how doctors can study their bodies.

decade -> 10 years
score -> 20 years

Summary

Part One: Life

We first meet Henrietta Lacks, a young African American mother of five, at Johns Hopkins Hospital on January 29, 1951. After experiencing pelvic pain and abnormal bleeding for more than a year, she and her husband drive nearly twenty miles from their home to the hospital in Baltimore, as it is one of the few locations at the time that treats black patients. Hopkins, which was built as a charity hospital for the sick and poor in 1889, is segregated; Henrietta sees Dr. Howard Jones, the gynecologist on duty, in a "colored-only" exam room. Dr. Jones finds a tumor on her cervix.

After her exam at Hopkins, Henrietta receives a diagnosis of stage I epidermoid carcinoma of the cervix. At the time of her diagnosis, Dr. Richard TeLinde, a Hopkins gynecologist and one of the leading experts on cervical cancer at the time, was researching new ways to detect and treat the disease. To test his hypotheses, he needs to study samples of cancerous cells from patients. He enlists George Gey, the intelligent, driven, yet slightly reckless, head of tissue research at the hospital, to help him harvest some of her cervical cells and put them under a microscope.

Meanwhile, Gey was working on his own project: trying to grow the first "immortal" human cells—ones that would continually reproduce themselves outside of the body. These cells could be used for any type of research on human cells that wouldn't be possible on a living person, such as finding a cure for cancer.

Over the course of Henrietta's treatment, surgeons remove a small piece of her normal cervical tissue and a piece from her tumor—without her knowledge or permission—and give them to Gey who, with the help of his lab assistant, Mary Kubicek, transfer the material into a culture for them to grow.

At the time, the medical community considers a patient's "participation" in medical research as a fair trade for free healthcare. While her healthy cells die soon after, her cancerous cells continue to duplicate every twenty-four hours. Amazed that he may have grown the first immortal human cells, Gey eagerly shares samples of Henrietta's cells with his colleagues. Before long, Henrietta's cells are in labs all over the world. Researchers can conduct experiments on them to see how human cells will react to difficult treatment methods.

While Henrietta bravely endures her treatments and continues to care for her children, she grows weaker and her pain intensifies. Doctors discover her cancer had spread throughout her body; little can be done, except try to ease her pain.

According to a colleague, Gey visits Henrietta in the hospital not long before her death and tells her that her cells will make her immortal. She replies that she is glad some good will come of her suffering. On October 4, 1951, less than a year after her diagnosis, Henrietta dies.

Before Henrietta's diagnosis, she had lived an average life. She was raised by her maternal grandfather, Tommy Lacks, in Clover, Virginia, where her ancestors had been slaves and the family still worked on tobacco farms. Henrietta grew close to her first cousin, David "Day" Lacks, who also lived with Tommy. Henrietta attended school through the sixth grade, while Day went through the fourth grade. Like all of the Lacks children, they left school young because their family needed their help in the fields.

Their relationship deepened and they had a son, Lawrence, and a daughter, Elsie, who was epileptic and had intellectual disabilities. Henrietta and Day married in August of 1941. After the bombing of Pearl Harbor just months later, the family moved to Turner Station, a small black community just outside Baltimore, so Day could take a job at Bethlehem Steel's Sparrows Point steel mill. With steel in great demand for the war effort, Day's new job was lucrative compared to his old work in the tobacco fields. Henrietta found friends in their new town and was loved by many. She was kind, fun, beautiful, caring, and always helped those in need.

In 1999, author Rebecca Skloot, who learned about the HeLa cells in a community college biology class, first tries reaching out to Henrietta's family. After an initial enthusiastic response from Henrietta's daughter Deborah, Skloot is stonewalled by Day Lacks and is stood up by Henrietta's son Sonny. Skloot also visits Clover and finds the once-bustling town is poverty-stricken, run down, and sparsely populated.

Need to Know: The book's first section gives us a glimpse of who Henrietta is as a person—information long ignored or forgotten by the scientific community that has worked with her cells for decades. As a wife and mother, Henrietta always puts her family first, even through her illness.

Racial prejudice is felt in every step of Henrietta's life and medical treatment—beginning with her first visit to Hopkins when she is segregated into the "colored-only" waiting and exam rooms. The treatment of African Americans by white doctors since the time of slavery and the resulting distrust of doctors and hospitals in Henrietta's family complicated their understanding of and feelings toward how her cells were used. They were resistant to talk to Rebecca Skloot, as journalists had long demanded to talk to them without ever offering any information on what happened to Henrietta's cells or why they were never informed or compensated.

Part Two: Death

After Henrietta's death, Day Lacks is left to care for his daughter, Deborah, and three sons, Lawrence, Sonny, and Joe (Elsie, who suffered from epilepsy, had been institutionalized before Henrietta's diagnosis). Poor and hungry, Henrietta's children have difficult childhoods. Deborah endures years of sexual abuse at the hands of her mother's cousin Galen, who has moved in with the family after Henrietta's death. His wife, Ethel, is equally cruel. She relentlessly beats and starves the children. Their father, Day, either doesn't believe his children, or looks the other way.

Lawrence drops out of high school to join the military. His girlfriend, Bobbette, becomes a mentor to the young Lacks children. She helps Deborah confront Galen's sexual abuse, and urges her to stay in school. Joe, Deborah's youngest brother and the sibling Ethel treats the worst (she often ties him up in the basement, beats him, and leaves him for long periods of time), grows up full of rage and is frequently in trouble with the law; in 1970, he is sent to prison for the murder of a neighborhood man. While in prison, Joe converts to Islam and changes his name to Zakariyya Bari Abdul Rahman.

Unbeknownst to her family, Henrietta's cells are on their way to becoming famous in the scientific community, because George Gey sends samples of her immortal cells to labs around the country. They are first used in the testing of Jonas Salk's polio vaccine, which helps stop one of the most widespread epidemics ever, and eradicates the disease in the United States.

Ironically, this study is led by the Tuskegee Institute in Tuskegee, Alabama, where, at the same time, horrific experiments are being conducted on black men with syphilis. In the now-notorious Tuskegee syphilis experiments, men are denied penicillin treatment so scientists can study the progression of the disease.

In 1953, a scientist in a Texas lab is running tests on the cells and sees chromosomes clearly for the first time, leading to the discovery that normal human cells have forty-six chromosomes. The cells are also used in commercial areas, including the cosmetics industry.

As more researchers want to use HeLa cells for more projects, the demand exceeds what the Tuskegee Institute can supply. Microbiological Associates, a company based in Bethesda, Maryland, creates a factory where the cells can be reproduced on a large scale; they sell to laboratories around the world. From the so-called "HeLa factory," the multibillion-dollar industry of selling human biological materials begins.

At the time, there is no clear ethical code for having patients or research subjects grant their consent for various tests or procedures, or for compensating them for any money made off their cells, tissues, or organs. For example, Chester Southam, a virology expert at Sloan Kettering Institute for Cancer Research, was curious to know if HeLa cells can infect healthy bodies.

He injected hundreds of healthy people with the cancer cells—mostly inmates from the Ohio State Penitentiary, without their knowledge—and found that healthy immune systems could fight off the disease. He also injected the cells into sick patients—without their consent—and found that the cells grew into tumors.

Some lawyers and doctors invoke the Nuremberg Code—drawn up in the aftermath of the horrific experiments Nazi doctors conducted on Jews during the Holocaust—which calls voluntary consent "absolutely essential," but the code is not law and it is difficult to ascertain how scientists and doctors are interpreting it. The government is nervous about "hindering scientific progress," so experimentation continues to be unregulated for the most part.

The phrase "informed consent" is first used in a legal document in 1957 after a patient sues his doctor when he becomes paralyzed from a routine surgery for which he was told there was no risk. Eventually, the National Institutes of Health requires that all projects they fund have to be examined by an independent board to ensure that consent by patients is given. However, there are hundreds of unethical experiments happening at the same time; Southam's was only one of many.

African Americans had even more reason to distrust the medical community, when it came to consent. Before the abolition of slavery, doctors tested drugs and performed operations on slaves without anesthesia, and in the early 1900s, corpses of black citizens were often exhumed from graves and sold to medical schools. From a young age Henrietta's children were reminded of stories they've heard about "night doctors" who kidnap black people and infect them with diseases.

When they finally learn about the HeLa cells, Day and his children are rightfully outraged that for so long they were not informed or compensated for the major role Henrietta's cells played in the development of twentieth-century medicine.

In 1960, HeLa cells make it to space. Scientists are studying how human cells act in orbit. They find that normal cells continue to divide normally, but cancerous cells such as HeLa cells become stronger, dividing at a much faster rate.

Geneticist Stanley Gartler makes a startling discovery in 1966 known as the "HeLa bomb." Henrietta's cells are discovered to be so potent, it is clear that scientists have been working with contaminated cell cultures in labs around the world, and the results of thousands of studies may have been tainted.

Gartler first infers this when he finds a rare genetic marker named G6PD-A that has only been detected in African Americans (and is present in HeLa), showing up in labs in numbers out of proportion to the general population. For this marker to turn up in cell cultures around the world means they must have been contaminated by HeLa.

Need to Know: In the decades after Henrietta Lacks's death, HeLa cells were used to develop the polio vaccine, study how cancer cells grow, and were sent up to space, all in the name of science. Despite the incredible success scientists were having, Henrietta's family was left struggling, financially—and was unaware that her cells were even taken.

The disconnect between her family and her cells reflect a medical institution that had not yet adopted a protocol for establishing patient consent and also had a legacy of racial discrimination.

Rebecca Skloot traces the early years of the patients' rights movement as some doctors and patients call for laws that make informed consent mandatory. Unfortunately, it is too late for Henrietta. Her family was shocked when they discovered that part of her was still alive, her cells were being sold to labs, and they received no compensation or recognition.

Part Three: Immortality

In 1973, a friend of the family—a doctor—passes through town. While catching up with Bobbette Cooper (née Lacks), he makes the connection that she is related to Henrietta Lacks—the source of the HeLa cells he used in his work. This is the first time the Lacks family hears that Henrietta's cells are alive and being used in labs throughout the world. Lawrence, Bobbette's husband, calls John Hopkins but no one is able to answer his questions.

After Stanley Gartler drops the HeLa bomb, scientists need to develop tests to isolate HeLa cells and stop any further contamination. To do so, researchers have to get genetic samples from Henrietta's relatives.

In the same year that the Lacks family learns about the HeLa cells, Susan Hsu, a postdoctoral fellow of John Hopkins geneticist Victor McKusick, comes to Day's house to take blood samples from his children. Hsu attempts to explain why she is interested in Henrietta's children's blood but between her broken English (Hsu is originally from China), her and Day's heavy accents, and Day's lack of education, the message is lost in translation. The family believes they are being tested for cancer.

After her earlier failed attempts to contact the Lacks family, Rebecca Skloot slowly begins to gain Deborah Lacks's confidence and she works with her to learn more about Henrietta and the HeLa cells. While the men in the Lacks family are more concerned with getting payment for HeLa cells—especially since selling the cells to labs has become a multibillion-dollar industry—Deborah is worried that she will develop cancer like her mother. Generations of the Lacks family struggled financially, and it is difficult for them to rationalize the profits from the HeLa business when they can barely afford basic necessities such as health insurance. When she never receives the results of what she believes to be her "cancer test" from Hsu's blood draw, Deborah becomes even more determined to learn about HeLa.

Together with Skloot, Deborah learns to interpret her mother's medical records and the articles about the HeLa cells that have long puzzled her. They travel to the site of what was once the Hospital for the Negro Insane, where Deborah's sister Elsie was institutionalized, and learn that she died at age fifteen.

Johns Hopkins cancer researcher Christoph Lengauer invites Skloot, Deborah, and Zakarriya—out of prison and in an assisted-living facility—to his lab to view HeLa cells through a microscope. For the first time, someone from Johns Hopkins answers their questions. The group learns that the reason Henrietta's cells are so strong is because of the nature of her tumor. HPV-18, a particularly aggressive strain of HPV, had inserted its DNA into her eleventh chromosome, turning off the gene that suppresses tumors. Ultra-strong cells contributed so much to science, from polio to AIDS research to discovering that only mutated cells are able to achieve immortality.

Zakarriya and Deborah are finally beginning to understand how their mother's cells contributed to science. The healing process has begun.

Deborah finds spiritual closure as well when she and Rebecca visit Henrietta's hometown of Clover and meet with her cousin Gary Lacks. A preacher, Gary performs a "soul cleansing," asking God to lift the burden of her mother's cells off of Deborah. Skloot is personally moved by the power of Deborah's faith, which has been able to soothe her fears and anger in a way that any answer from science could not.

Long suffering from high-blood pressure and diabetes, Deborah dies of a stroke in 2009.

Need to Know: Skloot brings Henrietta's story to the present day as she and Deborah examine the path HeLa cells took to become a tool that changed the course of medicine. Sadly, it took decades for anyone in the medical community to fully explain to the family why the cells were taken and how they were used.

Although her mother died when she was very young, Deborah devoted years of her life to learn what had happened to her mother's cells. The journey brought her years of fear and anxiety and affected her health. While her mother's biology helped scientists create new ways to diagnose and treat disease, Deborah struggled to pay for the medication she needed for her high-blood pressure and diabetes. Beyond understanding the science of what HeLa cells are and how they are used, Deborah needed a spiritual cleansing to finally feel closure. The religious aspect of Deborah's life provides a contrast to the hard science in the book—an answer from God was just as important to her than any explanation from scientists.

Timeline

1889: Johns Hopkins Hospital is founded.

1920: Henrietta Lacks is born in Roanoke, Virginia.

1947: The Nuremberg Code is created, the direct result of the medical atrocities conducted by Nazi doctors on prisoners during World War II. It is a set of ethical standards for human experimentation that requires informed patient consent.

1951: Henrietta Lacks dies of cervical cancer. George Gey successfully retrieves and grows the first immortal human cell line from Henrietta's cells. He names them HeLa.

1952: Scientists use HeLa cells to test and develop Jonas Salk's polio vaccine. The cells are supplied by the Tuskegee Institute, and later Microbiological Associates' "HeLa factory," which cultured HeLa cells and sold them to labs. HeLa cells become the first living cells to be shipped through the mail.

1953: HeLa cells are the first cells ever cloned.

1957: The term "informed consent" first appears in a court document when a patient sues his doctor after becoming paralyzed after a routine hospital procedure under anesthesia, which he was told posed no risk.

1966: Stanley Gartler drops the "HeLa bomb" at a conference after his discovery that HeLa cells may have contaminated other cell lines in labs around the world. The National Institutes of Health begins to require the approval of review boards to ensure patients have given consent for research they fund.

1971: Henrietta's son Joe is convicted of murder and sent to prison.

1973: More than twenty years after her death, Henrietta's family learns that her cells had been removed and are still alive. Researchers take blood samples from her progeny to further study HeLa cells, though they do so without explaining their purpose.

1976: *Rolling Stone* magazine publishes an article by Michael Rogers called "The Double-Edged Helix" that reveals the name and ethnicity of the woman from whom HeLa cells were harvested.

1988: Sixteen-year-old Rebecca Skloot first hears of Henrietta Lacks at a biology class and is fascinated by the missing pieces of her story.

2001: Deborah Lacks views HeLa cells for the first time under a microscope at Johns Hopkins.

2009: Deborah Lacks dies a year before the publication of *The Immortal Life of Henrietta Lacks.*

2013: Descendants of Henrietta Lacks come to an agreement with the National Institutes of Health with regard to sharing information about the HeLa genome and the family's right to privacy, while permitting researchers to gain access to the data.

Cast of Characters

Ethel and Galen: Henrietta and Day's cousins who stepped in to help Day with the children after Henrietta's death. This went terribly wrong. Ethel, who was always jealous of Henrietta because she believed that Galen preferred Henrietta to herself, physically and emotionally abused the children. Meanwhile, Galen sexually abused Deborah.

Stanley Gartler: Geneticist who dropped the "HeLa bomb" in 1966, uncovering his finding that HeLa cell samples around the world have been contaminated.

George Gey: Head of tissue culture research at Johns Hopkins who oversaw the growth of the first HeLa culture.

Susan Hsu: Postdoctoral fellow of Victor McKusick who took blood samples from Henrietta's children in 1973. She failed to adequately explain the purpose of her research and they believed they were being tested for cancer.

Howard Jones: The first doctor Henrietta saw at Johns Hopkins who diagnosed her cervical cancer and oversaw her treatment. It was Dr. Jones that definitively linked HeLa cells to Henrietta Lacks, confirming her identity in the journal *Obstetrics and Gynecology* in 1971.

Mary Kubicek: George Gey's laboratory assistant who created the first HeLa cell culture from which the immortal line grows.

Bobbette Lacks: Lawrence's wife. Bobbette became the Lacks children's mentor and surrogate mom after Henrietta's death. She convinced Deborah to stay in school and fend off her abusive cousin Galen's sexual advances. She also stopped Ethel from abusing the children, especially Joe. She is weary about the medical community and explains that many African Americans in Baltimore are, too, with the atrocious Tuskegee experiments to blame.

David "Day" Lacks: Henrietta's husband and the father of their five children. Also her first cousin, Day grew up with Henrietta in Clover, Virginia. Day was very suspicious of doctors and hospitals and of anyone who asked about the HeLa cells, including Skloot. He was unfaithful to Henrietta; he gave her syphilis and probably the HPV that led to cervical cancer. As a father, he was very hands-off. He ignored the sexual and physical abuse of his children at the hands of cousins Ethel and Galen. Skloot and Day's relationship is fairly strained.

David "Sonny" Lacks Jr.: Henrietta's third child. Sonny agreed at first to speak with Rebecca Skloot but he stood her up when she came to Turner Station to meet him. His heart is failing, but he can't afford healthcare.

Deborah "Dale" Lacks: Henrietta's fourth child. Deborah barely remembered her mother and she feared her cancer may be hereditary. Motivated to learn more about her mother and her own health, she agreed to work with Rebecca Skloot to uncover the story of the HeLa cells. She became Skloot's "guide" to the Lacks family until her death in 2009.

Elsie Lacks: Henrietta's second child. Elsie was epileptic and institutionalized at Crownsville State Hospital (formerly the Hospital for the Negro Insane) before Henrietta became ill. Elsie died at the hospital at age fifteen.

Henrietta Lacks: Grew up in rural Virginia. She left school in sixth grade to pick tobacco in her family's fields. She had five children with her cousin, and later husband, Day. At age thirty, she was diagnosed with cervical cancer and died soon after, in 1951. She was pretty, meticulous about her appearance, loved to go dancing and have fun, and was a loving wife and mother. It was from Henrietta that the immortal HeLa cells were harvested.

Lawrence Lacks: Henrietta's first child. Lawrence and his wife, Bobbette, helped care for his younger siblings after his mother's death. He is the leader of the Lacks family and, although suspicious about outsiders wanting to know about his mother, is proud of her contribution to making the world a better place.

Christoph Lengauer: Johns Hopkins cancer researcher who reached out to Deborah and Zakariyya and invited them to his lab where he answered their questions about HeLa cells and showed them the cells under a microscope. After a lifetime of being ignored by the science community, Lengauer's compassionate attempt to honor Henrietta's family helped them heal.

Victor McKusick: Geneticist at Johns Hopkins who performed research on samples taken from Henrietta's children without their knowledge or consent.

McKusick hoped to discover what made HeLa cells unique so they could be detected in contaminated samples.

Zakariyya Bari Abdul Rahman (born Joe Lacks): Henrietta's fifth and youngest child. As a very young boy, after his mother's death, he was brutally abused by his cousins. He grew up into an angry man, who was convicted of murder and sentenced to fifteen years of prison in 1971. He converted to Islam and changed his name while incarcerated. After his release, he had difficulty holding down a job and drinks. A moment of healing occurs at Christoph Lengauer's lab, however, as he starts to understand how incredible his mother's contribution to the world really was.

Rebecca Skloot: Author of the book, Skloot becomes a character herself in *The Immortal Life of Henrietta Lacks* as she works with Deborah Lacks to learn more about Henrietta's life and HeLa cells.

Chester Southam: An immunologist and oncologist at Sloan Kettering Cancer Center and Cornell University Medical Center. Curious to know if HeLa cells could infect the scientists that were handling them, Southam conducted clinical research on patients without their informed consent by injecting them with cancerous HeLa cells. Tumors grew on those that already had cancer and therefore weak immune systems. He also injected the cells into healthy patients—inmates who gave consent—but their strong immune systems successfully fought off the cancer. After hundreds of people had already been injected without their consent, three Jewish doctors, aware of the Nuremburg Code, refused to carry out his experiments. Southam went on to become the president of the American Association of Cancer Research, despite his history of questionable ethics.

Richard Wesley TeLinde: One of the leading experts on cervical cancer in Henrietta's time and a doctor at Johns Hopkins. For his research into the diagnosis and treatment of the disease, he takes cell samples from many women, including Henrietta.

Direct Quotes and Analysis

"When I saw those toenails . . . I nearly fainted. I thought, Oh Jeez, she's a real person. *I started imagining her sitting in her bathroom painting those toenails, and it hit me for the first time that those cells we'd been working with all this time and sending all over the world, they came from a live woman. I'd never thought of it that way."*

In this quote, Mary Kubicek, George Gey's lab assistant who created the first HeLa culture, tells Rebecca Skloot what it was like to see Henrietta's body after her death. That Kubicek was only able to imagine Henrietta as a "real person" after seeing her painted toenails reveals the disconnect between the medical community and the people from whom they harvested the cells and tissues used in their research.

"If the whole profession is doing it, how can you call it 'unprofessional conduct'?"

Spoken by a lawyer defending American virologist Chester Southam, the man who injected prisoners with HeLa cells to see if they would develop cancer. When Southam proposed continuing his study on patients at the Jewish Chronic Disease Hospital in Brooklyn in 1963, the hospital's doctors opposed exposing patients to cancer cells without their knowledge or consent. They cited the Nuremberg Code that forbade such experimentation after the horror of Nazi doctors performing tests on Jewish prisoners. William Hyman, a member of the hospital's board of directors, led the charge against Southam.

But Southam's lawyer's defense reveals how widespread and accepted this type of research was in the medical community, even a decade after Henrietta's death.

"You'd be surprised how many people disappeared in East Baltimore when I was a girl. . . . I lived here in the fifties when they got Henrietta, and we weren't allowed to go anywhere near Hopkins. When it got dark and

we were young, we had to be on the steps, *or Hopkins might get us.*"

Bobbette Lacks, Henrietta's daughter-in-law, tells Rebecca Skloot about the belief prevalent in African American communities in Baltimore when she was younger that Johns Hopkins doctors would kidnap or kill black people for scientific research. Their fears are not without basis, though, as there is evidence that doctors had experimented on black slaves without consent, and, around the turn of the twentieth century, corpses of black people were frequently stolen by grave robbers and sold to medical schools. As evident in the Tuskegee Syphilis Experiment that ran from 1932 to 1972, in which black men were denied treatment for syphilis so scientists could study the natural progression for the disease, the black community has reason to distrust scientists.

"It was the first time anyone had told the true story of Henrietta Lacks and her family, the first time the mainstream media had reported that the woman behind HeLa was black. The timing was explosive."

Here, Rebecca Skloot refers to the 1976 *Rolling Stone* article that reveals the identity and racial original of the HeLa cells used in labs around the world. She provides useful historic context, in that the ethics of the Tuskegee syphilis study had already been called into question, and this was another example of a mostly white industry being perceived as taking advantage of people of color, using their cells—and their bodies—for medical research without consent. On the other side of this exploitation is the notion that in the 1970s, there were "Archie Bunkers" who may have been shocked to learn that their medical treatment may have been derived from a black woman who descended from slaves.

"I can't say nuthin bad about science, but I won't lie, I would *like some health insurance so I don't got to pay all that money every month for drugs my mother cells probably helped make."*

This quote from Deborah Lacks illustrates the conflict that members of the Lacks family felt with the medical establishment. Great advancements have been made in medicine, thanks to Henrietta's cells, but her own children lived in poverty and Deborah struggled to afford the medicine for her high-blood pressure and diabetes. Deborah poses one of the book's central questions: Should individuals be compensated for the use of their cells or tissues?

"As Zakariyya and Christoph walked away, [Deborah] raised the vial and touched it to her lips. 'You're famous,' she whispered. 'Just nobody knows it.'"

When Johns Hopkins cancer researcher Christoph Lengauer brought Deborah and Zakariyya to his lab to look at HeLa cells under a microscope and answered their questions about their mother's cells, Deborah finally felt a moment of peace and closure. For years, she's been terrified that part of her mother was suffering from the experiments being done on HeLa cells, and sad that scientists around the world had access to the mother she could barely remember.

Trivia

1. Henrietta Lacks's cells were found to contain an especially potent strain of HPV (human papillomavirus)—one of the factors that helped them live forever. Because of research done on HeLa cells, scientists have been able to develop a test and vaccine for the virus, saving women from the cervical cancer that took Henrietta's life.

2. HeLa cells have been flown into space on missions led by both Russia and the United States to study how human cells behave in zero gravity.

3. Jonas Salk developed the polio vaccine using HeLa cells. Thanks to the vaccine, the deadly disease has been eradicated in the United States and much of the world.

4. Combined, all the HeLa cells produced in the world would weigh a total of 50 million metric tons and, if stacked end-to-end, they would span more than 350 million feet.

5. Henrietta's story has inspired at least two songs: "The Cells That Will Not Die" by Jello Biafra and the Guantanamo School of Medicine, and "Henrietta" by Yeasayer.

6. There is a public school in Vancouver, Washington, named Henrietta Lacks Health and Bioscience High School. Locals refer to it as "HeLa High."

7. HeLa cells were used in atomic bomb testing to discover the effects nuclear bombs would have on humans.

8. Oprah Winfrey and HBO Films are producing a film based on *The Immortal Life of Henrietta Lacks* and began shooting in the summer of 2016. The cast includes Courtney B. Vance, Renée Elise

Goldsberry, and Leslie Uggams; Winfrey plays Deborah Lacks.

9. HeLa cells have inspired several works of art. In *HeLa on Zhora's coat*, artist Aleksandra Domanović transformed images of HeLa cells into a pattern on a raincoat. *HeLa*, an installation by Christine Borland, features a petri dish filled with HeLa cells under a microscope with live images of the cells multiplying in the dish projected on a screen.

10. Rebecca Skloot created the Henrietta Lacks Foundation, a nonprofit organization that provides financial assistance to individuals in need who have made contributions to science without their knowledge or consent. Grants have been given to the Lacks family and descendants of research subjects used in the Tuskegee syphilis studies. A portion of the profits from sales of her book are contributed to the foundation.

What's That Word?

Cell: The smallest unit of biological life that can reproduce independently.

Cell culture: The process through which extracted cells are grown outside of their natural environment—typically in labs.

Cell line: A population of cells descended from a single cell; they contain identical genetic makeup.

Chromosome: Within the nucleus of a cell, it is the structure made up of protein and DNA that contains genetic material. Humans have forty-six chromosomes. HeLa cells helped scientists understand the human chromosome, a major breakthrough.

DNA: Deoxyribonucleic acid, DNA, is a molecule that holds the genetic instructions for all living organisms and some viruses, including growth, development, and reproduction. The HPV virus has inserted its DNA into one of Henrietta Lacks's chromosomes, turning off the gene that suppresses tumors.

G6PD-A (glucose-6-phosphate dehydrogenase-A): A rare genetic marker found nearly exclusively in African Americans—and in HeLa cells. When G6PD-A showed up in samples in labs well out of proportion than what is in the population, Stanley Gartner was able to determine they had been contaminated by HeLa.

HeLa: The HeLa cell line, derived from cervical cells taken from Henrietta Lacks, is the oldest immortal cell line and remains the most commonly used in labs around the world.

HPV (human papillomavirus): The most common sexually transmitted disease in the United States. Henrietta Lacks's cells were found to contain a virulent strain of HPV, which contributed to their immortality. Research on HeLa cells has led to a vaccine for HPV.

Informed consent: A legal procedure to ensure that a patient knows all the risks and the costs involved in a treatment or a procedure, or before he or she enrolls in a clinical trial. Henrietta Lacks did not give consent when a doctor removed pieces of tissue from her cervix.

Critical Response

- A *New York Times* bestseller
- An Ambassador Book Award in American Studies winner
- An American Association for the Advancement of Science's Award for Excellence in Science Writing winner
- A *Chicago Tribune* Heartland Prize for Nonfiction winner
- An Indie Lit Award for Nonfiction winner
- A Medical Journalists' Association Open Book Award winner
- A Wellcome Trust Book Prize winner

"[A] multilayered story about 'faith, science, journalism, and grace.' It is also a tale of medical wonders and medical arrogance, racism, poverty . . . [A] rich, resonant tale of modern science, the wonders it can perform and how easily it can exploit society's most vulnerable people."

—*Publishers Weekly* (starred review)

"Just as the DNA in a cell's nucleus contains the blueprint for an entire organism, so does the story of Henrietta Lacks hold within it the history of medicine and race in America, a history combining equal parts of shame and wonder.. . . The role Skloot assumed when she stepped into this narrative could serve as a textbook example for the mission of narrative journalists everywhere." —Laura Miller, *Salon*

"[Skloot's] absorbing book is not just about medicine and science but about colour, race, class, superstition and enlightenment, about the painful, transfixing romance of being American." —Hilary Mantel, *The Guardian*

"Rebecca Skloot introduces us to the 'real live woman,' the children who survived her, and the interplay of race, poverty, science and one of the most important medical discoveries of the last 100 years. Skloot narrates the science lucidly, tracks the racial politics of medicine thoughtfully and tells the Lacks family's often painful history with grace."

—*The New York Times*

"Skloot weaves a fascinating and tender detective story about HeLa's legacy through the discovery of Henrietta's youngest daughter, Deborah, who didn't know her mother but who always knew she wanted to be a scientist. As Skloot and Deborah, infinitely different yet united by the shared quest for answers, unravel one of the most absorbing mysteries of modern science, we also get a rich and sensitive tale about family, community, and the dark side of society's capacity for exploiting its poorest and most vulnerable members."

—Brain Pickings

"Equal parts intimate biography and brutal clinical reportage, Skloot's graceful narrative adeptly navigates the wrenching Lack family recollections and the sobering, overarching realities of poverty and pre–civil-rights racism. The author's style is matched by a methodical scientific rigor and manifest expertise in the field."

—*Kirkus Reviews* (starred review)

"Writing with a novelist's artistry, a biologist's expertise, and the zeal of an investigative reporter, Skloot tells a truly astonishing story of racism and poverty, science and conscience, spirituality and family driven by a galvanizing inquiry into the sanctity of the body and the very nature of the life force."

—*Booklist* (starred review)

"This extraordinary account shows us that miracle workers, believers, and con artists populate hospitals as well as churches, and that even a science writer may find herself playing a central role in someone else's mythology" —*The New Yorker*

"[A] remarkable feat of investigative journalism and a moving work of narrative nonfiction that reads with the vividness and urgency of fiction. It also raises sometimes uncomfortable questions with no clear-cut answers." —*National Public Radio*

About Rebecca Skloot

Born in Springfield, Illinois, Rebecca Skloot was raised in the Pacific Northwest. She worked as a veterinary technician for ten years before going on to earn her BS in biological sciences at Colorado State University and an MFA in creative writing from the University of Pittsburgh.

As a science journalist, she has contributed to the *New York Times*, the *New York Times Magazine*, *Prevention*, *O: The Oprah Magazine*, *New York Magazine*, *Popular Science*, NPR's *Radiolab*, and PBS's *NOVA ScienceNOW*. Her work has been included in the Best Food Writing and the Best Creative Nonfiction anthologies. The story of Henrietta Lacks fascinated her since she first heard about HeLa cells in a community college course she was sitting in on as a teenager. Published in 2010, *The Immortal Life of Henrietta Lacks* is her first book.

For Your Information

Online

“10 Unapologetic Books About Race in America.” EarlyBirdBooks.com

“Deal Done Over HeLa Cell Line.” Nature.com

“A Family Consents to a Medical Gift, 62 Years Later.” NYTimes.com

“Five Reasons Henrietta Lacks Is the Most Important Woman in Medical History.” PopularScience.com

“‘Henrietta Lacks’: A Donor’s Immortal Legacy.” NPR.com

“Henrietta Lacks’ ‘Immortal’ Cells.” Smithsonian.com

“How Rebecca Skloot Built *The Immortal Life of Henrietta Lacks*.” TheOpenNotebook.com

LacksFamily.net

"Learning the Wrong Lessons on Privacy from Henrietta Lacks." WashingtonPost.com

"Your Cells. Their Research. Your Permission?" NYTimes.com

Books

Bad Blood: The Tuskegee Syphilis Experiment by James H. Jones

Culturing Life: How Cells Became Technology by Hannah Landecker

The Emperor of All Maladies: A Biography of Cancer by Siddhartha Mukherjee

In Search of Our Mothers' Gardens by Alice Walker

Medical Apartheid: The Dark History of Medical Experimentation on Black Americans from Colonial Times to the Present by Harriet A. Washington

Mountains Beyond Mountains: The Quest of Dr. Paul Farmer, a Man Who Would Cure the World by Tracy Kidder

Paid Servant by E. R. Braithwaite

Bibliography

Debatty, Régine. "Bodily Matters: Human Biomatter in Art (Part 5: Working with HeLa Cells, Microflora and Other Biomedical Material)." *We Make Money Not Art.* August 17, 2016. http://we-make-money-not-art.com/bodily-matters-human-biomatter-in-art-part-5-working-with-hela-cells-microflora-and-other-biomedical-material/.

Etchells, Pete. "Jonas Salk Google Doodle: A Good Reminder of the Power of Vaccines." *The Guardian.* October 28, 2014. https://www.theguardian.com/science/head-quarters/2014/oct/28/jonas-salk-google-doodle-celebrates-polio-vaccine-pioneer.

Goodwyn, Tom. "Yeasayer Reveal New Track 'Henrietta.'" *NME.* May 16, 2012. http://www.nme.com/news/music/yeasayer-9-1268827.

Henrietta Lacks Foundation. "About the Henrietta Lacks Foundation." http://henriettalacksfoundation.org/.

Kamen, Jess. "Holiday in Baltimore." *Baltimore City Paper.* June 23, 2014. http://www.citypaper.com/music/bcp-cms-1-1707767-migrated-story-cp-2014-06-25musi-20140623-story.html.

Laufe, Anne. "New Vancouver High School Will Focus on Health and Medical Careers." *The Oregonian.* October 2, 2012. http://www.oregonlive.com/clark-county/index.ssf/2012/10/new_vancouver_high_school_will.html.

Nodell, Bobbi. "UW Researchers Report on Genome of Aggressive Cervical Cancer That Killed Henrietta Lacks." *UW Today*, University of Washington. August 7, 2013. http://www.washington.edu/news/2013/08/07/uw-researchers-report-on-genome-of-aggressive-cervical-cancer-that-killed-henrietta-lacks/.

Pederson, Erik. "Courtney B. Vance, Leslie Uggams, 7 Others Join Oprah Winfrey in HBO's 'Immortal Life of Henrietta Lacks.'" *Deadline.* August 11, 2016. http://deadline.com/2016/08/courtney-b-vance-leslie-uggams-immortal-life-of-henrietta-lacks-oprah-winfrey-hbo-1201801925/.

Rogers, Michael. "The Double-Edged Helix." *Roll-*

ing Stone. March 25, 1976. http://www.rolling-stone.com/culture/features/the-double-edged-helix-19760325.

MORE SMART SUMMARIES

FROM WORTH BOOKS

POPULAR SCIENCE

MORE SMART SUMMARIES

FROM WORTH BOOKS

EMPOWERMENT

WORTH BOOKS

SMART SUMMARIES

MORE SMART SUMMARIES

FROM WORTH BOOKS

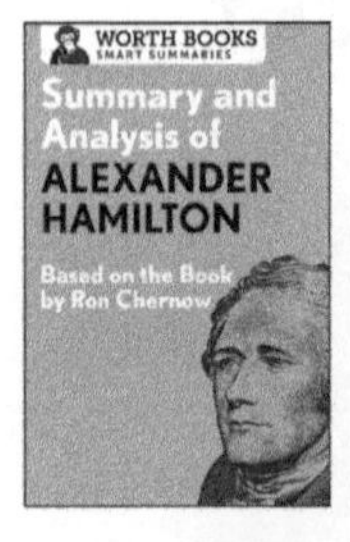

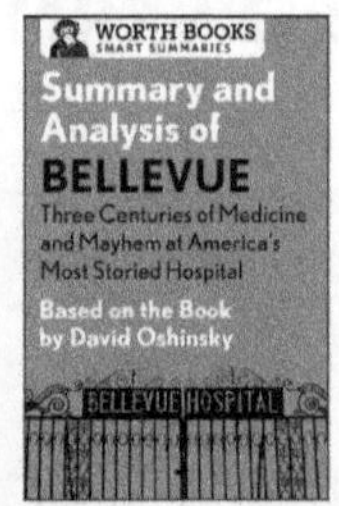

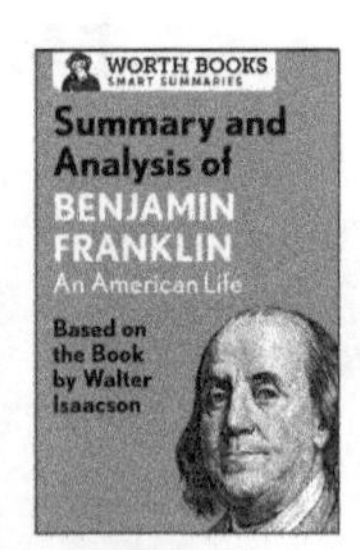

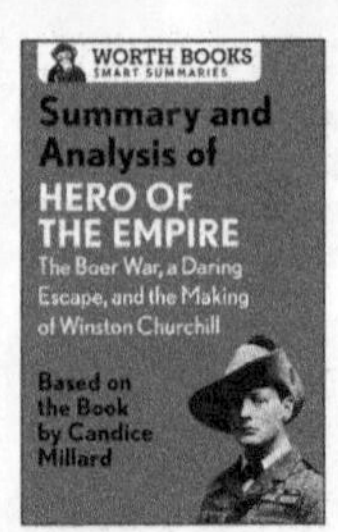

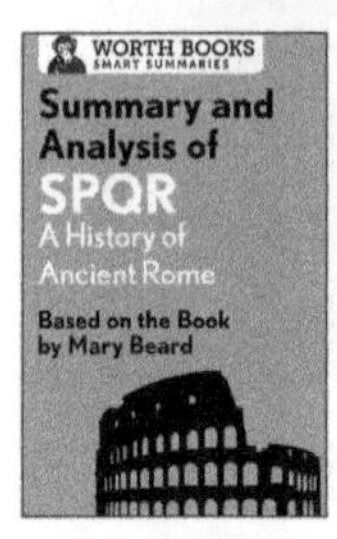

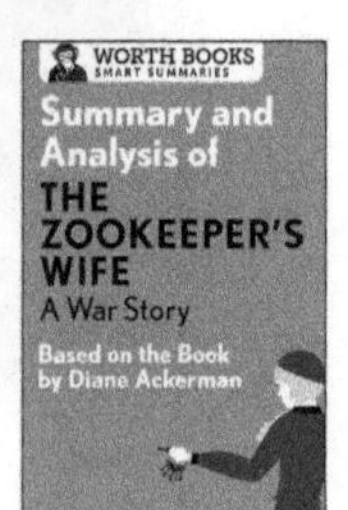

WORTH BOOKS

SMART SUMMARIES

OPEN ROAD
INTEGRATED MEDIA

CPSIA information can be obtained
at www.ICGtesting.com
Printed in the USA
LVHW021458300520
656909LV00006B/1278